TOO CUTE! BABY ANIMALS

BABY BACKYARD ANIMALS

RACHAEL MORLOCK

PowerKiDS press

Published in 2025 by The Rosen Publishing Group, Inc.
2544 Clinton Street, Buffalo, NY 14224

First Edition

Editor: Rachael Morlock
Book Design: Michael Flynn

Photo Credits: Cover, p. 1 Luis Enrique Torres/Shutterstock.com; (series background) Viktoriia LiSa/Shutterstock.com; p. 5 Dobbography/Shutterstock.com; p. 7 Mike Truchon/Shutterstock.com; p. 9 Breck P. Kent/Shutterstock.com; p. 11 Alexey Stiop/Shutterstock.com; p. 13 Mak Mak/Shutterstock.com; p. 15 Margaret M Stewart/Shutterstock.com; p. 17 Debbie Steinhausser/Shutterstock.com; p. 19 Betty Shelton/Shutterstock.com; p. 21 Amy Lutz/Shutterstock.com.

Library of Congress Cataloging-in-Publication Data

Names: Morlock, Rachael, author.
Title: Baby backyard animals / Rachael Morlock.
Description: [Buffalo, New York] : PowerKids Press, [2025] | Series: Too cute! Baby animals | Includes bibliographical references and index.
Identifiers: LCCN 2024003947 (print) | LCCN 2024003948 (ebook) | ISBN 9781499445145 (library binding) | ISBN 9781499445138 (paperback) | ISBN 9781499445152 (ebook)
Subjects: LCSH: Garden animals–Juvenile literature. | Urban animals–Juvenile literature. | Garden animals–Infancy–Juvenile literature. | Urban animals–Infancy–Juvenile literature. | Garden animals–Habitat–Juvenile literature. | Urban animals–Habitat–Juvenile literature.
Classification: LCC QL119 .M67 2025 (print) | LCC QL119 (ebook) | DDC 591.75/6–dc23/eng/20240305
LC record available at https://lccn.loc.gov/2024003947
LC ebook record available at https://lccn.loc.gov/2024003948

Manufactured in the United States of America
Some of the images in this book illustrate individuals who are models. The depictions do not imply actual situations or events.

CPSIA Compliance Information: Batch #CSPK25. For Further Information contact Rosen Publishing at 1-800-237-9932.

CONTENTS

Animal Neighbors

No matter where you live, wild animals and their babies live nearby. You might see geese in the park, squirrels in your yard, raccoons near your trash bin, or rabbits in your garden. You can find the animals in this book in the city and the countryside!

Robin Hatchlings

Baby robins **hatch** from pale blue eggs in a nest. The parents keep them safe. The parents feed the hungry hatchlings worms and bugs. Over two weeks, the babies grow feathers and become as big as their parents. Then they're ready to leave the nest!

Hawk Chicks

In the spring, red-tailed hawk **mates** build a high nest. The mother lays up to five eggs, and both parents keep them warm. After they hatch, the father hunts for food. He finds small **mammals**, birds, and **reptiles**. The mother feeds small pieces to the chicks.

Cottontail Kits

Many backyard rabbits are cottontails. They're born in grass-covered nests on the ground. Newborn rabbits are called kittens or kits. After a few days, kits grow fur, open their eyes, and become cuter than ever! They drink their mother's milk for about a month.

Squirrel Kits

Mother tree squirrels build nests for their babies in tree branches or hollows. Like rabbits, newborn squirrel kits are **bald** and **blind**! They don't open their eyes for about four weeks. Kits count on their mother to feed them her milk and keep them safe.

Chipmunk Pups

Chipmunks are the smallest kind of squirrel. A mother chipmunk usually makes her nest in an underground hole called a burrow. She has a **litter** of two to eight pups in the spring or summer. A newborn chipmunk is not much bigger than a bumblebee!

Skunk Kits

Baby skunks are stinkin' cute! They're born in an underground burrow in the spring. A mother skunk can have 12 kits in a litter. Kits stay in the burrow for about six weeks. Then, they follow their mother out at night to find food.

Raccoon Cubs

Like skunks, raccoons are **nocturnal**. Their babies are called cubs or kits. A few cubs are born together. They spend their first ten weeks in a den, which could be a hole in a tree or log. Cubs live with their mother for over a year.

Keeping Watch

You can keep watch for baby animals in your backyard or neighborhood park in the spring. They often stay hidden in nests or burrows while they're very small. You might not see them until they're a few weeks old. By then, they'll be cuter than ever!

GLOSSARY

bald: Hairless.

blind: Without sight.

hatch: To break open or come out of an egg.

litter: A group of babies born to one animal at the same time.

mammal: Any warm-blooded animal whose babies drink milk and whose body is covered with hair or fur.

mate: One of two animals that come together to produce babies.

nocturnal: Active mainly during the night.

reptile: A cold-blooded animal with thin, dry pieces of skin called scales.

FOR MORE INFORMATION

BOOKS

Munson, Victoria. *Urban Wildlife*. London, UK: Wayland, 2020.

Rodger, Ellen. *Animals on the Outskirts*. St. Catherines, ON: Crabtree Publishing Company, 2020.

Zobel, Derek. *Squirrels*. Minneapolis, MN: Bellwether Media, 2021.

WEBSITES

American Robin
www.audubon.org/field-guide/bird/american-robin
Learn more about the American robin and how it raises its young.

Chipmunk
kids.nationalgeographic.com/animals/mammals/facts/chipmunk
Read this description of the chipmunk to learn where it lives and what it eats!

Publisher's note to educators and parents: Our editors have carefully reviewed these websites to ensure that they are suitable for students. Many websites change frequently, however, and we cannot guarantee that a site's future contents will continue to meet our high standards of quality and educational value. Be advised that students should be closely supervised whenever they access the internet.

INDEX